Gaia Ierace

Dracula and his Family

商務印書館

CONTENTS

Originally published by Black Cat Publishing under the title: *Dracula and his Family*
© 1999 Black Cat Publishing
An imprint of Cideb Editrice, Genoa, Canterbury

The copyright of this Chinese edition is owned by
The Commercial Press (H.K.) Ltd.

Name of Book: Dracula and his Family
Author: Gaia Ierace
Editors: Monika Marszewska, Elvira Poggi Repetto
Design and Art Direction: Nadia Maestri
Illustrations: Claudio Decataldo
Layout: Sara Blasigh

系 列 名：Quality English Learning for Kids · I
書　　名：Dracula and his Family
責任編輯：傅　伊
出　　版：商務印書館（香港）有限公司
　　　　　香港筲箕灣耀興道 3 號東滙廣場 8 樓
　　　　　http://www.commercialpress.com.hk
印　　刷：美雅印刷製本有限公司
　　　　　九龍觀塘榮業街 6 號海濱工業大廈 4 樓 A
版　　次：2004 年 9 月第 1 版第 1 次印刷
　　　　　© 2004 商務印書館（香港）有限公司
　　　　　ISBN 962 07 1722 8
　　　　　Printed in Hong Kong

出版說明

　　學英語當然要學優質的，有品質才能讓人有信心。我們一直積極提倡學習優質英語的理念，並且為學習者提供過多元化的優質英語材料，像《Black Cat 優質英語階梯閱讀》就十分成功，至今已出版近 60 本。鑑於良好的英語能力最好從小培養，我們於是出版這一套適合五至八歲兒童的優質英語閱讀讀本 "Quality English Learning for Kids・I"。

　　培養兒童對於英語的興趣，須從趣味和簡易兩方面入手。圖文並茂，聲文結合這兩大特點對學習英語甚有幫助。"Quality English Learning for Kids・I" 承續本館出版優質英語書的理念，全書彩圖精美，附 CD 朗讀內容及聆聽練習，形式多元化，有出版故事讀本（story books）、圖畫讀本（picture readers）、戲劇讀本（drama readers）及互動讀物（interactive readers）四大類，提供不同學習功能。故事讀本和圖畫讀本可供兒童看圖講故事；戲劇讀本完全用對白編寫，培養脫口而出講英語的習慣，適合家庭裏作簡單的角色扮演，或者小學生在課堂上作簡單的演出。

　　針對兒童學習英語的需要，本系列提示家長為兒童設定學習目標，並且說明如何達標，另備生詞表和語法知識點，讓兒童在家長協助下掌握生詞用法，認識簡單的句子結構和了解語法要點。

　　"Quality English Learning for Kids・I" 吸引兒童對閱讀產生興趣，逐步引導他們參與愉快的閱讀旅程。在這個旅程中，家長是重要的導航者，透過對兒童的悉心鼓勵，循循善誘，進一步加強親子關係。

商務印書館
編輯部

使用説明

如何使用本書？

本書為戲劇讀本（drama reader），適合課堂使用或親子共讀。

每頁均圖文並茂。正文為對話，讓小孩子熟悉並掌握第一、二、三身句法。老師或家長可讓小孩子用第一身敘述句法作自我介紹，用第二身疑問句法提問，用第三身敘述句法描述某人或某物。老師或家長還可與小孩子扮演劇中角色進行對話，練習使用第一、二、三身句法。

除正文外，還設有練習題（選擇、填色、連線、配對等），培養小孩子的辨別和記憶能力。老師或家長可指出近旁實物，讓小孩子説出物體的名稱、顏色和形狀。

本書配有CD，小孩子可邊聽邊讀，提高英語聽説能力。

本書的學習目標是甚麼？

老師或家長可為孩子定出以下學習目標。

使用本書後，孩子學會：

(a) 自我介紹（introduce myself）；

(b) 説出某人喜歡的事物（describe what someone likes）；

(c) 辨別家庭成員（identify family members）；

(d) 聽從CD的指示，玩連線（join the dots）、搭配（listen and match）、填色（listen and colour）等遊戲。

本書有哪些重點生詞和語法學習項？

(a) 重點生詞：本書的重點生詞包括五大類，即房屋（house）、家庭（family）、食品（food）、服裝（clothes）和日常活動（daily routine）。另附英漢對照生詞表，增強對生詞的理解和記憶。

(b) 語法學習項：

第一身句法（the first person）（例如頁4，"I'm Count Dracula's butler." 和頁42，"I like his neck!"）

第二身句法（the second person）（例如頁6，"Are you there Alex?" 和頁9，"You're welcome!"）

第三身句法（the third person）（例如頁7，"He's in his coffin." 和 "They have coffins."）

形容詞的比較級（comparatives），結構為 "形容詞＋-er"（adj+-er）。（例如頁18，"My bone is bigger." 和頁45，"Every day he got fatter and fatter!"）

由can構成的疑問句（questions formed with *can*），用於向某人詢問某事或徵求某人對某事的意見，can置於主語之前。（例如頁7，"Can you see him?" 和頁24，"Can I have some more blood, dad?"）

現在進行時（present continuous），動詞形式為 "動詞原形＋-ing"（base form +-ing）。（例如頁7，"Count Dracula is sleeping." 和頁13，"Brian's crying."）

PICTIONARY 1

butler

castle

coffin

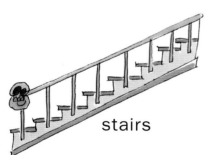

stairs

blood

follow me

bat

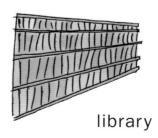

library

3

CHAPTER 1

Count Dracula's Castle

 Roman: Hello! My name's Roman Blenski. I'm Count Dracula's butler. And this is Count Dracula's castle.

Roman: Come in!

WELCOME TO
TRANSYLVANIA.
MY CASTLE IS
YOUR HOME.
YOURS, VLAD.

Roman: Follow me up the stairs and down again.
  Are you there Alex?
Alex: Yes, I am.
Roman: And you, Joanna?
Joanna: Yes, I am.
Roman: Good.

Roman: Sshhhh! Count Dracula is sleeping. Can you see him?

Alex and Joanna: Yes, we can.

Roman: He's in his coffin. Vampires don't have beds. They have coffins. What colour is it?

Alex: It's black...

Joanna: ...and white.

Roman: That's right. It's black and white.

Roman: Here is Count Dracula's library. The Count loves books. Can you see them?

Alex: Yes, strange books!

Joanna: They're very strange! *Bats and Vampires; Blood, my life.*

QUALITY ENGLISH CLUB
Membership Application Form

QUALITY ENGLISH CLUB is for those who love English reading and seek for better English to share and learn with fun together.

Benefits offered:
- *Membership Card*
- *English learning activities*
- *English learning e-forum*
- *Surprise gift and more...*

Simply fill out the application form below and fax it back to 2565 1113 or send it back to the address at the back.

Join Now! It's FREE exclusively for readers who have purchased books on Quality English Learning published by the Commercial Press!

(Please fill out the form with **BLOCK LETTERS**.)

The title of book(s) /book set(s) that you have purchased: _____

English Name: _____ (Surname) _____ (Given Name)

Chinese Name: _____

Address:

Tel: _____ Fax: _____

Email: _____

Sex: ❏ Male   ❏ Female   (Login password for e-forum will be sent to this email address.)

Education Background: ❏ Kindergarten   ❏ Primary 1-3   ❏ Primary 4-6
❏ Junior Secondary Education (F1-3)   ❏ Senior Secondary Education (F4-5)
❏ Matriculation   ❏ College   ❏ University or above

Age: ❏ 3 - 6   ❏ 6 - 9   ❏ 10 - 12   ❏ 13 - 15   ❏ 16 - 18
❏ 19 - 24   ❏ 25 - 34   ❏ 35 - 44   ❏ 45 - 54   ❏ 55 or above

Occupation: ❏ Student   ❏ Teacher   ❏ White Collar   ❏ Blue Collar
❏ Professional   ❏ Manager   ❏ Business Owner   ❏ Housewife
❏ Others (please specify: _____)

As a member, what would you like **QUALITY ENGLISH CLUB** to offer:
❏ Member gathering/ party   ❏ English class with native teacher   ❏ English competition
❏ Newsletter   ❏ Online sharing   ❏ Book fair
❏ Book discount   ❏ Others (please specify: _____)

Other suggestions to **QUALITY ENGLISH CLUB**: _____
_____

Please sign here: _____   (Date: _____)

Visit us at Quality English Learning Online http://publish.commercialpress.com.hk/qel

QUALITY ENGLISH CLUB

The Commercial Press (Hong Kong) Ltd.
8/F, Eastern Central Plaza,
3 Yiu Hing Road, Shau Kei Wan,
Hong Kong

THE COMMERCIAL PRESS (H.K.) LTD.

Roman: This is Count Dracula's desk. The Count writes letters to his friends in Scotland, Ireland, Russia and Bavaria. This is our pet, Pat. Pat the bat!

Joanna and Alex: Hi, Pat!

Roman: I must go now.

Alex and Joanna: Thank you very much, Mr Blenski.

Roman: You're welcome!

9

ACTIVITY A

Choose the correct answer.

1 Count Dracula's castle is in:
- **a.** Scotland.
- **b.** Transylvania.
- **c.** Russia.

2 Dracula's coffin is:
- **a.** pink.
- **b.** red.
- **c.** black and white.

3 Dracula loves:
- **a.** butlers.
- **b.** friends.
- **c.** books.

4 Dracula has friends in:
- **a.** Russia.
- **b.** Italy.
- **c.** France.

5 Dracula has a pet. His name is:
- **a.** Tommy.
- **b.** Pet.
- **c.** Pat.

ACTIVITY B

Join the dots to draw Dracula's castle.

ACTIVITY C

Act out the story. Ask your friends / classmates to help you.

A: You are Roman.
B: You are Alex.
C: You are Joanna.

PICTIONARY 2

bone

Blood card

thin

receipt

big - bigger

fangs

pint of blood

CHAPTER 2

Home sweet home

Adam: Stop it Eve!

Eve: It's my bone, not yours!

Vlad: Brian's crying.

Felicia: Shhh, my poor baby!

Vlad: Felicia! It's late.

Felicia: OK. Let's go to the blood supermarket!

Blood assistant: Good morning, Count and Countess.
Can I help you?

Vlad: A pint of blood, please.

Blood assistant: Just one pint?

Felicia: Two pints, please.
Little Brian is very thin.

Blood assistant: Anything else?

Felicia: Two pints of blood squash for Adam and Eve
and a Vamp lollipop for Brian.

Blood assistant: Here you are, Countess.

Two pints of blood 43p
Two pints of blood
squash 55p
A Vamp lollipop 12p

Vlad: How much is that?

Blood assistant: It's
1 pound 10 pence, sir.

Vlad: Here's my Blood
card.

Blood assistant:
Thank you sir.
Here's your
card and your
receipt.

Vlad: Thank you.
Goodbye.

Felicia: Goodbye.

Vlad: OK, Roman. Let's go to the zoo!
Roman: Yes, sir.

Eve: What's that, mum?
Felicia: It's a Vamp bike. Do you like it?
Eve: Yes, I do. It's very nice.

Adam: What's that dad?
Vlad: It's Fast Fangs restaurant.

Roman: Here we are, sir.

Felicia: Come on kids. Let's go!

Adam: Let's play with the bats!

Eve: Yes, let's go!

Felicia: What are you reading, dear?

Vlad: The *Vampire Times.*

Felicia: Come on Brian! Good boy!

Adam: Come here, bat! Here's a delicious bone for
  you.
Eve: My bone is bigger! Come here!
Adam: Roman, where are
  my other bones?
Roman: Here they are.
Adam: Thank you.

Felicia: It's time to go, kids.
Eve and Adam: Oh no, dad!
Vlad: Come on!

ACTIVITY A

Choose the correct answer.

1 Felicia and Vlad have:
   a. two children.
   b. three children.
   c. four children.

2 Felicia and Vlad go to:
   a. the blood supermarket.
   b. the bank.
   c. the museum.

3 Felicia buys blood for Brian because he is very:
   a. old.
   b. thin.
   c. happy.

4 Vlad pays:
   a. 1 pound.
   b. 1 pound 17 pence.
   c. 1 pound 10 pence.

5 At the zoo Adam and Eve play with the:
   a. bats.
   b. cats.
   c. dogs.

19

ACTIVITY B

Listen and fill in the family tree.

Dracula's Family Tree

ACTIVITY C

Act out the story. Ask your friends / parents / classmates to help you.

A: You are Adam.

B: You are Eve.

C: You are Vlad.

D: You are Felicia.

E: You are the Blood assistant.

F: You are Roman.

PICTIONARY 3

gloves

hat

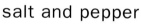

salt and pepper

son

daughter

soup

chips

CHAPTER 3

Dinner at Vlad's castle

5

Vlad: Are you ready, dear?
Felicia: Yes, just a moment. Where are my gloves?
Vlad: On the table.
Felicia: Thank you.

Vlad: Where's my hat?
Felicia: It's in the yellow box.
Vlad: Great! Let's go downstairs!

Vlad: Good evening, Mr and Mrs Dark. Welcome to my castle!

Mr Dark: Good evening, Count and Countess. This is my son Jack and this is my daughter Blackie.

Jack and Blackie: Hello!

Felicia: This is Adam and this is Eve.

Adam and Eve: Good evening!

Roman: Dinner is ready, sir.

Vlad: Thank you, Roman.

Eve: Can I have some more blood, dad?

Vlad: Of course, dear.

Mrs Dark: The soup is lovely!

Felicia: Thank you. It's Roman's special dish.

Mr Dark: Salt and pepper?

Jack: Yes, please.

Blackie: What's that, Mrs Dracula?

Felicia: This is a Vampburger and chips and that's
   black ice cream. Delicious!

Roman: Coffee, sir?
Mr Dark: Yes, please.
Roman: Sugar?
Mrs Dark: One, please.
Vlad: What about you, Felicia?
Felicia: Two, please.

Mr Dark: You've got a wonderful collection of coffins, Count!
Vlad: Thank you! This is a Roman coffin and that's Russian.

Adam: This is my fang.

Blackie: No, it isn't!

Eve: It's *my* fang!

Adam: No, it's our grandad's fang!

Eve: Our grandad Igor is a very strong man. He's a famous champion.

Jack: Is he?

Adam: Yes. Look, here's his photo.

Jack and Blackie: Wow!

Adam: Come in!

Mr Dark: Time to go, kids!

Blackie and Jack: OK, dad.
  This is our vamp card,
  Adam.

Adam: Thanks.

Blackie and Jack: Bye!

Adam and Eve: Bye-bye!

Mr and Mrs Dark: Thank you very much,
  Count Dracula. Thank you, Countess.

Vlad and Felicia: You're welcome.
  Goodbye!

ACTIVITY A

Choose the correct answer.

1 Felicia's gloves are:
- **a.** under the table.
- **b.** in the box.
- **c.** on the table.

2 Mr and Mrs Dark have:
- **a.** 3 children: Jack, Jill and Lucy.
- **b.** two children: Jack and Blackie.
- **c.** two children: John and Tom.

3 The ice cream is:
- **a.** red.
- **b.** blue.
- **c.** black.

4 Vlad has a collection of:
- **a.** bones.
- **b.** coffins.
- **c.** stars.

5 Igor is Adam and Eve's:
- **a.** grandad.
- **b.** dad.
- **c.** mum.

ACTIVITY B

Listen and colour the boxes with the right colour.

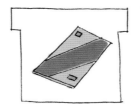

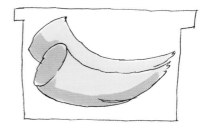

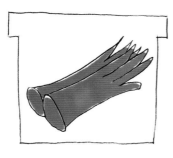

ACTIVITY C

Act out the story. Ask your friends / parents / classmates to help you.

A: You are Vlad.

B: You are Felicia.

C: You are Mr Dark.

D: You are Jack.

E: You are Blackie.

F: You are Adam.

G: You are Eve.

H: You are Roman.

I: You are Mrs Dark.

PICTIONARY 4

chair

stripes

jumper

size: small / medium / large

stars

shoes

flowers

milk

dress

strawberry

CHAPTER 4

The family goes shopping

Felicia: Come on, kids!

Eve: Just a moment, mum. Where's my hat?

Adam: It's over there. On the chair.

Eve: Thanks, Adam.

Roman: The car's ready, sir.

Vlad: Thank you, Roman.

Shop assistant: Can I help you, madam?

Felicia: My son likes that jumper.

Shop assistant: The yellow one with blue stripes, madam?

Adam: Yes!

Shop assistant: Here you are!

Adam: Thank you. Can I try it on, please?

Shop assistant: Yes, of course.

Felicia: Do you like it, Vlad?

Vlad: It's too big! What size is it?

Shop assistant: It's a large.

Felicia: Can we have a small?

Shop assistant: Of course, madam. Here you are.

Adam: But ... this is orange with white stripes! I like yellow.

Shop assistant: What about this jumper? It's yellow with blue stars.

Adam: Great!

Eve: Can I have a dress, mum?

Felicia: Which dress do you like?

Eve: The red dress
with flowers.

Felicia: Do you like it,
Vlad?

Vlad: Yes, I do.

Eve: Can I try it on, please?

Shop assistant: Yes. Here you are!

Eve: I like it!

Vlad: Yes, it's very nice.

Felicia: Look! Do you like these shoes, Eve?

Eve: Yes, I do.

Vlad: They're perfect for the dress.

Eve: Can I have them
    dad, please?

Vlad: OK, dear.

Eve: Thanks dad,
    thanks!

Vlad: Do you like your ice cream, Eve?

Eve: Yes, it's lovely.

Adam: What flavour is it?

Eve: It's chocolate. And yours?

Adam: It's strawberry and vanilla.

Vlad: How's your coffee, Felicia?

Felicia: It's very hot.

Can I have some milk, please?

Vlad: Yes.

Here you are!

36

ACTIVITY A
Choose the correct answer.

1 The family goes:
    a. to the cinema.
    b. to the restaurant.
    c. shopping.

2 Adam likes:
    a. the red jumper.
    b. the blue dress.
    c. the yellow jumper.

3 Adam's jumper is too:
    a. small.
    b. big.
    c. expensive.

4 Eve likes:
    a. the red jumper.
    b. the blue dress.
    c. the red dress.

5 At the coffee shop the children have:
    a. hamburgers.
    b. ice cream.
    c. chips.

ACTIVITY B

Listen and match. Now colour the pictures.

I like...

blue

red

black

yellow

pink

ACTIVITY C

Act out the story. Ask your friends / parents / classmates to help you.

A: You are Felicia.

B: You are Eve.

C: You are Adam.

D: You are Roman.

E: You are Vlad.

F: You are the shop assistant.

G: You are the second shop assistant.

PICTIONARY 5

garlic

dark

a bite

midnight

fat - fatter

cold

dreams

map

neck

moon

cross

CHAPTER 5

Vlad and Felicia go out

Felicia: ...and the monster opens its mouth and has a wonderful dinner.
Vlad: Blood soup?
Felicia: Yes, dear.

Vlad: It's midnight. Goodnight, kids!
Adam and Eve: Goodnight, dad. Goodnight, mum.
Felicia: Goodnight. Sweet dreams!

Vlad: Are you ready, dear?

Felicia: Just a moment. Where's my map?

Vlad: Here it is.

Felicia: Thank you. Let's visit the homes with a red cross.

Vlad: Yes, OK. What about the black crosses?

Felicia: Black is for GARLIC! These people like garlic very much.

Vlad: Garlic!!! NO!!!

Felicia: Calm down! Come on, Pat. Let's go!

Pat: Yeeesss.

Felicia: It's cold and dark. It's lovely!
Pat: It's a dark night! No moon, no stars!
Vlad: Perfect!

Felicia: What about this woman?
Vlad: Oh no, she's too thin.
Pat: Yes, too thin!

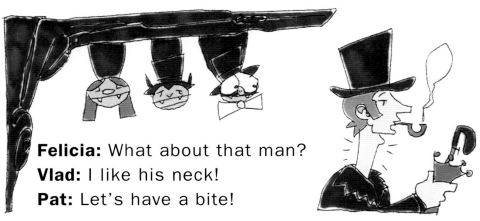

Felicia: What about that man?
Vlad: I like his neck!
Pat: Let's have a bite!

Vlad: Mmm! Delicious!
Felicia: Yes, very good!
Pat: Wonderful!

Felicia: Let's go to the blood bank. I need a pint of blood for little Brian. Have you got your Blood card?
Vlad: Yes, here it is. There's a blood bank in Horror Street.

Felicia: What's your code, dear?
Vlad: 660XX.
Felicia: 660XX.

Welcome to Horror Street Blood Bank, Madam! Can I help you?

Felicia: A pint of blood, please.
Blood Bank Machine: Press the red button, please.
Felicia: Thank you.

Pat: I'm so tired!

Vlad: Can I have a glass of whisky, Roman?

Roman: Of course, sir! Here's some Transylvanian whisky.

Vlad: Thank you.

Felicia: Once upon a time there was a vampire named Vlad.

He lived in a big castle and he loved blood.

Blood for breakfast, blood for lunch, blood for supper...

Every day he got fatter and fatter!

ACTIVITY A

Choose the correct answer.

1 Vlad and Felicia go out:
- **a.** in the morning.
- **b.** at night.
- **c.** in the afternoon.

2 Felicia's map has:
- **a.** red and white crosses.
- **b.** blue crosses.
- **c.** red and black crosses.

3 Felicia, Vlad and Pat go to the blood bank in:
- **a.** Horror Street.
- **b.** Oxford Street.
- **c.** Vamp street.

4 Vlad's secret code is:
- **a.** XXOIU.
- **b.** XXO66.
- **c.** 66OXX.

5 Vlad likes:
- **a.** whisky.
- **b.** water.
- **c.** orange juice.

ACTIVITY B

Listen and colour.

ACTIVITY C

Act out the story. Ask your friends / parents / classmates to help you.

A: You are Felicia.

B: You are Vlad.

C: You are Adam.

D: You are Eve.

E: You are Pat.

F: You are the Blood Bank Machine.

G: You are Roman.

47

Dracula's world: a night rhyme

Match each picture with the right lines.

**Dracula likes the night.
At night he has a bite!**

*Felicia goes shopping
for Adam and Eve:
a jumper, a dress
and some blood if you please!*

**The Draculas love bats –
not cats!**

*Their pet, Pat
is a wonderful bat!
Pat wears a black hat
and a bow tie.
Goodbye!*

DRACULA AND HIS FAMILY

Page 10 – Activity A

1b   2c   3c   4a   5c

Page 11 – Activity B

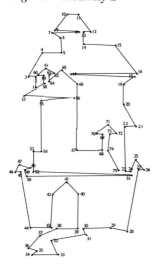

Page 11 – Activity C

Open activity.

Page 19 – Activity A

1b   2a   3b   4c   5a

Page 20 – Activity B

Listening:
Hello! I'm Eve and this is my family.
This is my mum, Felicia, and this is my
dad, Vlad.
I've got two brothers, Adam and Brian.
Adam is ten years old and
Brian is two years old. I'm eight years old.
We've got a pet. His name is Pat. He is a
lovely bat. He's three years old. We love
Pat very much.

Page 20 – Activity C

Open activity.

Page 28 – Activity A

1c   2b   3c   4b   5a

Page 29 – ActivityB

Listening:
The Blood card is in the green box.
The Vamp-bike is in the blue box.
The bat is in the yellow box.
The hat is in the red box.
Felicia's gloves are in the black box.
Igor's fangs are in the grey box.

49

Page 29 – Activity C

Open activity.

Page 37 – Activity A

1c   2c   3b   4c   5b

Page 38 – Activity B

Listening:
Hello! I'm Adam. I like yellow jumpers.
Good afternoon! I'm Roman. I like pink shoes.
Hi! I'm Felicia. I like blue gloves.
Hello! I'm Eve. I like red dresses.
Good evening! I'm Vlad. I like black hats.

Page 46 – Activity A

1b   2c   3a   4c   5a

Page 47 – Activity B

Listening
Colour the hat for Dracula BLACK.
Colour the chocolate ice-cream BROWN.
Colour the shoes for Roman PINK.
Colour the dress for Eve RED.
Colour the jumper for Adam YELLOW and BLUE.
Colour the Vamp lollipop GREEN and ORANGE.

Page 47 – Activity C

Open activity.

Page 48 – DRACULA'S WORLD: A NIGHT RHYME

Dracula likes the night.
At night he has a bite!

Felicia goes shopping
for Adam and Eve:
a jumper, a dress
and some blood if you please!

The Draculas love bats –
not cats!

Their pet, Pat
is a wonderful bat!
Pat wears a black hat
and a bow tie.
Goodbye!

GLOSSARY

bite 咬

blood bank 血庫

bow tie 蝴蝶結

breakfast 早餐

butler 男管家

button 按鈕

champion 冠軍

cinema 電影院

coffin 棺材

count 伯爵

countess 伯爵夫人

delicious 美味的

dinner 晚宴

expensive 昂貴的

famous 著名的

fang 尖牙

flavour 味道

goes shopping 去買東西

hamburgers 漢堡包（麵包夾煎牛肉餅）

library 圖書館

lollipop 棒棒糖

lovely 可口的

lunch 午餐

orange juice 橘子汁

pepper 胡椒

pet 寵物

press 按，壓

receipt 收據

restaurant 飯館

salt 鹽

size（衣服）尺碼

shop assistant 售貨員

squash 果汁汽水

strange 奇怪的

sugar 糖

supermarket 超級市場

supper 晚餐

vampires 吸血鬼

vanilla 香草（冰淇淋）

whisky 威士忌酒

zoo 動物園